Arthur J. Lembo, Jr.
2015

First Printing: June, 2015

Arthur J. Lembo, Jr.
1440 East Sandy Acres Drive
Salisbury, MD 21804

www.artlembo.wordpress.com

Preface

In 2004, I created a small publication with my students titled *How Do I Do that In ArcGIS/Manifold.* It was an enormously fun endeavor, which to my surprise really took off in the GIS community. There were tens of thousands of downloads, some lengthy debate, and finally some attempts to replicate the document for other GIS software products. Seeing the document take on a life of its own was really gratifying. This current document is a continuation of that theme, but with a focus on the use of spatial SQL to accomplish the tasks.

Once again, we will revisit the 1988 United States Geological Survey (USGS) classic document titled *The Process for Selecting Geographic Information Systems*[1] (Guptil, et. al., 1988). As you might recall from the previous *How Do I Do That* document, the USGS report provided an overview of the process for selecting geographic information systems, in addition to a checklist of functions that a GIS should include. The functions were broken into five separate categories: user interface, database management, database creation, data manipulation and analysis, and data display and presentation.

This book is the first of three publications on how to achieve the USGS tasks using the SQL syntax. While this book illustrates the GIS processes allowable in Manifold GIS 8.0, a companion book includes a discussion of PostGRES/PostGIS. Soon, I expect to release the third book on using the SQL engine under development by Manifold Software, Ltd. (currently named Radian).

As the title indicates, I envision this book to act as a sort of reference to the question of *how do I do that..*, residing on the user's lap while attempting to implement GIS functionality with SQL. Also, I believe simply perusing through the page will be convincing enough to cause users to consider the use of SQL as part of their daily GIS activities. Also, as a teaching tool, one can see how so many of the same SQL functions are just reused in a different fashion to complete a task, without the need to run some kind of special wizard or new tool. So, if you can write SQL, you can build all kinds of functionality in GIS.

You will also notice that this book is much shorter than the original *How do I do that in ArcGIS/Manifold*, and without illustrations. This was done on purpose so as to keep the cost of production low and to allow users to quickly get an answer to their spatial SQL questions. Finally, all of the examples have been tested using a sample Manifold 8 .map project file, downloadable from my blog. Therefore, users can try to recreate the SQL in the book on the actual data. To do that of course, the queries in this book must be retyped. This too was done on

---

[1] Guptill, S., D. Cotter, R. Gibson, R. Liston, H. Tom, T. Trainer, H. VanWyhe. 1988. "A Process for Selecting Geographic Information Systems". Technology Working Group – Technical Report 1. USGS Open File Report 88-105.

purpose - retyping the queries is the best way to learn. Simply copying and pasting the queries is not going to help you learn what is actually happening. So, while there is a little more effort on the part of the reader, I believe it will be the most effective way to learn how to write SQL. As you become more accustomed to writing SQL, you will find that you begin to *think in SQL*. For me, when presented with a GIS conundrum, I constantly find myself thinking about the SQL solution, rather than the classical GIS commands.

SQL is a very easy to understand query language, even for those who are unfamiliar with programming. When I began this book, I had no idea whether Manifold 8 could complete a majority of the tasks using only the SQL engine. I was quite pleased to see that virtually all of the tasks listed in *How do I do that in ArcGIS/Manifold* could actually be accomplished with the SQL engine. And if that is the most you get out of this book, I think the effort was worth it.

I welcome you to participate in the discussion of this book on my blog: **artlembo.wordpress.com**.

Arthur J. Lembo, Jr.
June, 2015

Table of Contents

How to understand this guide

This guide follows the topic headings from the book *How do I do that in ArcGIS/Manifold*, as a way to illustrate the capabilities of Manifold's SQL engine for accomplishing classic GIS tasks. You will notice that several gaps exist where the tasks cannot be completed using SQL, and would require the GUI. In some instances the gaps are logical as the specified task requires user interaction. In other cases, however, the gap exists because a function that one might assume would be a logical addition to the software simply was not built into the GIS engine. A good example of this is the creation of area features from connected (spaghetti) lines.

All of the SQL queries were tested on a Manifold .map file named "how do I - Manifold 8.map" that can be downloaded from artlembo.wordpress.com. Although generally similar to ANSI SQL, the syntax is designed for the Manifold 8 SQL engine and has some features particular to the software.

Field names are always surrounded by brackets ([ ]) when a space exists in either the field name or table name, for example [parcel table]. While it is good practice to utilize brackets for fields that do not have spaces (for example [hydro]), for brevity, some of the examples do not utilize brackets.

Manifold allows access to geometry values in a table through either the [ID] field, or the [Geom (I)] field. In most instances, it is easier to simply type ID, however for clarity this document uses the [Geom (I)] so that the reader understands that a geometry field is being used.

I have also attempted to indent the queries as best as possible so that it is easier to read. However, there are some examples that have too long of names, or too much code to fit on a single line on the page. In these cases, there may be some minor *wrapping* of the text.

Database Management

Database management functions provide for tracking, retrieval, storage, update, protection, and archiving of stored data.

page 29. The Process for Selecting Geographic Information Systems

Adding a column to a table

Adding a column to a table is relatively straightforward in Manifold SQL. The original *How do I* document did not include options like changing the name, data type, or size of the columns. Nonetheless, the following shows a few examples on how to modify tables using the `ALTER TABLE` command in SQL:

```
ALTER TABLE uses the following syntax:

ALTER TABLE table {
ADD [COLUMN] column type [(size)] [DEFAULT default] |
 ALTER [COLUMN] column type [(size)] [DEFAULT default] |
ALTER [COLUMN] column SET DEFAULT default |
DROP [COLUMN] column |
RENAME [COLUMN] column TO column
}
```

Add a column

To add a new column (in this case, a column named **HomeAge** that stores Integers, the user simply writes:

```
ALTER TABLE [parcels]
ADD COLUMN [HomeAge] INTEGER
```

Rename a column

```
ALTER TABLE [parcels]
RENAME COLUMN [HomeAge] TO [Age_of_Home]
```

Change the column type

```
ALTER TABLE [parcels]
ALTER COLUMN [propclass] CHAR
```

Remove a column

```
ALTER TABLE [parcels]
DROP COLUMN [Age_of_Home]
```

Sorting tabular or graphical data

Using SQL, we can sort the data on any field, combination of fields, or even an on-the-fly calculated field. Sorting can be done in either ascending or descending order:

```
ORDER BY expression [ASC | DESC] [, expression [ASC | DESC] ...]
```

Sort in descending order

```
SELECT [parcelkey], [asmt] FROM parcels
ORDER BY [asmt] DESC
```

Sort by multiple fields

```
SELECT parcelkey, acres, asmt, addrstreet, FROM parcels
ORDER BY asmt, acres
```

Sort based on an on-the-fly calculation[2]:

```
SELECT NAME, (asmt - land) AS structurevalue
FROM parcels
ORDER BY structurevalue DESC
```

[2] when renaming a column from an SQL query, we typically use the AS directive. However, Manifold 8 does not require that you use AS - although its inclusion allows for easier reading

Calculating values for new fields using arithmetic or related tables - making field calculations.

New values may be calculated using the `UPDATE` statement, or values can be calculated on-the-fly without changing the actual data in a table. You can calculate data as a result of an SQL `SELECT` query, or as a calculation into an existing field. Generally, calculating new values is performed using the `UPDATE` statement on a column for a table:

```
UPDATE table SET column = {expression | DEFAULT} [, column =
{expression | DEFAULT} ...] [WHERE ...]
```

The following example updates a field named [homevalue] in the parcels table by subtracting the value of the land from the entire assessment value [asmt] of the property:

```
UPDATE parcels
SET homevalue =  parcels.[asmt] - parcels.[land]
```

<u>Calculations on a related table</u>

You can also use a table relation, and update the value in one table with values in another table that are related by a common field. In this example we are subtracting the land value in the parvalues table from the asmt value in the parcels table. To accomplish this, the inner SELECT query performs a calculation on the fly, and also relates the parcel table to the parvalues table. This inner query is treated as a virtual table that is then used within the UPDATE statement:

```
UPDATE
 (SELECT parcels.[asmt] - parvalues.[land] AS calc_homevalue
  FROM parcels, parvalues
  WHERE parcels.[parcelkey] = parvalues.[parcelkey]
 )
SET homevalue = calc_homevalue
```

<u>Calculations without updating the table</u>

The following performs a calculation on-the-fly to determine the tax amount for a property based upon its assessed value as part of a `SELECT` query without updating an existing field:

```
SELECT (parcels.[asmt] * .07) AS TaxValue
FROM parcels
```

Relating data files and fields

SQL relates can work on more than one table, as this example illustrates how to relate multiple tables together based on a related item. While many complex relates are possible in Manifold SQL, simple tables are used here illustrate this process.

There are many options for relating data files in SQL through a Join function, as defined by:

```
TABLE {LEFT | RIGHT | FULL} [OUTER] JOIN TABLE ON <condition>
```

For example, we can join two tables (parcels and propclas) as:

```
SELECT parcels.[parcelkey],
parcels.[propclass],propclas.[description]
FROM parcels, propclas
WHERE parcels.[propclass] = propclas.[value]
```

Unfortunately, the previous query will return an error because the [value] field in the propclas table is represented as a text value, while the [propclass] field in the parcels table is an Integer. However, SQL can create the relationship by simply changing the data type of the [value] field on-the-fly to an Integer cy converting the text to an Integer (CInt):

```
SELECT parcels.[parcelkey],
parcels.[propclass],propclas.[description]
FROM parcels, propclas
WHERE parcels.[propclass] = CInt(propclas.[value])
```

We can also join the tables on the left or right using the `JOIN`[3] statement:

```
SELECT parcels.[parcelkey], parcels.[propclass],
propclas.description
FROM parcels
RIGHT JOIN propclas
ON parcels.[propclass] = CInt(propclas.[value])
```

Right outer joins include all of the records from the second (right) of two tables, even if there are no matching values for records in the first (left) table.

---

[3] The JOIN statement is ordinarily a faster method for joining data than the full scan approach shown previous to it. However, Manifold 8 seems to handle both in the same amount of time.

```
SELECT parcels.[parcelkey], parcels.[propclass],
       propclas.description
FROM parcels
LEFT JOIN propclas
ON parcels.[propclass] = CInt(propclas.[value])
```

Left outer joins include all of the records from the first (left) of two tables, even if there are no matching values for records in the second (right) table.

Database Creation

Database creation functions are those functions required to convert spatial data into a digital form that can be used by a GIS. This includes digitizing features found on printed maps or aerial photographs and transformation of existing digital data into the internal format of a given GIS.

Page 29, The Process for Selecting Geographic Information Systems

Digitizing

Ordinarily, digitizing is performed within the GUI of a GIS where the user points-and-clicks on the screen. However, sometimes SQL can be used if you have a table of coordinate values you want to place in, or if you are receiving input from say an Internet based application. The following examples illustrate how to insert geometries into an existing layer. For simplicity, we will restrict our data to latitude and longitude, but other coordinate systems can be used.

Points are created from two coordinate values. To insert a point in latitude and longitude, we simply write:

```
INSERT INTO parcels ([Geom (I)])
Values (NewPointLatLon(-76.5,42.45))
```

If you had a table of latitude and longitude values, you could insert all of them into a new layer as:

```
INSERT into trees ([geom (i)])
 (SELECT NewPointLatLon(lng,lat)
  FROM treetable
)
```

Lines are created from a series of point geometries. To add a Line, you could write:

```
INSERT INTO hydro ([Geom (I)])
Values (NewLine(NewPointLatLon(-76.5,42.44),
                NewPointLatLon(-76.58,42.48)
                )
        )
```

Areas are created as a series of coordinates[4]. To add a an Area, you could write:

```
INSERT INTO parks ([Geom (I)])
Values(ConvertToArea(
                     NewLine(NewPointLatLon(-76.5,42.45),
                     NewPointLatLon(-76.58,42.45),
                     NewPointLatLon(-76.58,42.9)
                             )
                     )
         )
```

[4] You don't have to close the area on itself in Manifold 8 - the ConvertToArea function assumes that you want the area closed and will snap the last point you enter to the first point in the geometry.

Assigning Topology - identifying intersection points

You can find intersection points for either line or area geometries in a single layer as:

```
SELECT IntersectionPoint(roads.[Geom (I)],roads.[Geom (I)])
FROM roads
```

Or, you can identify intersection points of multiple vector layers as:

```
SELECT IntersectionPoint(Roads.[Geom (I)],hydro.[Geom (I)])
FROM Roads,hydro
WHERE Intersects(hydro.[Geom (I)],Roads.[Geom (I)])
```

Also, you can insert the intersection points into a new layer using the `INSERT` command:

```
INSERT into scratchlayer ([geom (I)])
  (SELECT IntersectionPoint(Roads.[Geom (I)],hydro.[Geom (I)])
   FROM Roads,hydro
   WHERE Intersects(hydro.[Geom (I)],Roads.[Geom (I)])
  )
```

Creating a Polygon from Line Segments (not illustrated)

Creating a polygon from line segments in Manifold 8 is performed in the GUI and is not built into the SQL engine.

Creating a distance buffer from line segments

Buffers can be created on any type of geometry, either points, lines, or areas - using the Buffer statement as:

```
Buffer(geometry,distance,units)
```

The units can be virtually any unit such as meters, feet, miles, kilometers, etc.

Buffer with a constant value

```
SELECT Buffer([Geom (I)],100,"m")
FROM [hydro]
```

Buffer with an attribute assigned value

```
SELECT Buffer([Geom (I)],[BufferDist],"m")
FROM [hydro]
```

Creating ringed buffers around a geometry[5]:

```
SELECT buffer([Geom (I)], 50) AS g
FROM [parks]

UNION ALL
SELECT buffer([Geom (I)], 30) AS g
FROM [parks]

UNION ALL
SELECT buffer([Geom (I)], 20) AS g
FROM [parks]
```

[5] UNION ALL assumes that the table structure for each query is the same - therefore, you must have the same fields and field types.

Correcting topological errors - eliminating overlaps, undershoots, and dangles (not illustrated)

Correcting topological errors in Manifold 8 is performed in the GUI and is not built into the SQL engine.

Import and export - importing database tables, raster data, and vector data (not illustrated)

Importing and exporting data in Manifold GIS is a function of the GUI, and therefore not built into the SQL engine.

Data Manipulation and Analysis

Data manipulation and analysis functions provide the capability to selectively retrieve, transform, restructure, and analyze data.

Retrieval options provide the ability to retrieve either graphic features or feature attributes in a variety of ways. Transformation includes both coordinate/projection transformations and coordinate adjustments. Data restructuring includes the ability to convert vector data to raster data, merge data, compress data, reclassify or rescale data, and contour, triangulate, or grid random or uniformly spaced z-value data sets

Analysis functions differ somewhat depending on whether the internal data structure is raster or vector based. Analysis functions provide the capability to create new maps and related descriptive statistics by reclassifying and combining existing data categories in a variety of ways. Analysis functions also support: replacement of cell values with neighboring cell characteristics (neighborhood analysis); defining distance buffers around points, lines and areas (proximity analysis); optimum path or route selection (network analysis); and generating slope, aspect and profile maps (terrain analysis).

Page 29, The Process for Selecting Geographic Information Systems.

Vector Retrieval - Select by Window

Although *select by window* normally assumes an interactive session with the GUI, you can use SQL to selecting by a window in a similar fashion to selecting by a bounding box. In this example, a box is created by entering a centerpoint, buffering the centerpoint by 50 units, and creating a box around the buffer. All vector features touching the box are selected. Instead of touching, the query could use contains or intersects:

```
OPTIONS CoordSys("parcels" AS COMPONENT);
SELECT * FROM parcels
WHERE touches(parcels.[Geom(I)],
              BoundingBox(Buffer(AssignCoordSys
 (newpoint(841891,890701),CoordSys("parcels" AS Component)),500)))
```

We can also create a polygon to select the features that touch:

```
SELECT * FROM parcels
WHERE touches(parcels.[Geom (I)],
(ConvertToArea(NewLine(NewPointLatLon(-76.5,42.45),
                       NewPointLatLon(-76.58,42.45),
                       NewPointLatLon(-76.58,42.9)
                             )
                     )
                )
             )
```

In the above example, the polygon is simply a triangle. You could easily add more points to create a more detailed polygon, or you could create a rectangle.

Raster Retrieval - select data by area masks

Raster pixels may be selected using an existing vector layer by updating the pixel's [Selection (I)] value. Because raster pixels aren't a geometry, so you first have to convert each pixel to a geometry (in our example, we are using NewPointLatLon to convert the pixel coordinates to latitude and longitude):

```
OPTIONS CoordSys("dem" AS Component);
UPDATE (
   SELECT dem.*
   FROM dem, parks
   WHERE touches(parks.[geom(i)],
         NewPointLatLon([dem].[Longitude (I)],
                        [dem].[latitude (i)]))
       )
SET [Selection (i)] = true
```

Or, you can create a polygon on the fly like our earlier examples:

```
OPTIONS CoordSys("dem" AS Component);
UPDATE (
  SELECT dem.*
  FROM dem, parks
  WHERE touches((ConvertToArea(NewLine
                          (NewPointLatLon(-76.25,42.45),
                           NewPointLatLon(-76.58,42.45),
                           NewPointLatLon(-76.58,42.9)
                                      )
                                     )
                                )
                ,NewPointLatLon([dem].[Longitude (I)],
                 [dem].[latitude (i)])
                )
        )
SET [Selection (i)] = true
```

Data restructuring - convert from raster to vector; and vector to raster (not illustrated)

This task is done as part of the GUI, and not supported in the Manifold SQL engine.

Modify raster cell size by resampling (not illustrated)

This task is done as part of the GUI, and not supported in the Manifold SQL engine.

Reducing unnecessary coordinates – weeding (not illustrated)

Weeding coordinates in Manifold GIS is a function of the GUI, and therefore not built into the SQL engine.

Smoothing data to recover sinuosity (not illustrated)

Smoothing data in Manifold GIS is a function of the GUI, and therefore not built into the SQL engine.

Data restructuring - changing raster values by selected area or feature

One of the nice things about understanding SQL is how it can build on previous concepts. Previously we used SQL to select pixels that were within a window, or touching a geometry. One that task is completed, you can simply wrap the result into an UPDATE statement, and instead up updating the [Selection (I)] value in the raster surface, you can update the [Height (i)] field as follows:

```
OPTIONS CoordSys("empty_dem" AS Component);
UPDATE
(
  SELECT empty_dem.*
  FROM empty_dem
  WHERE touches(ConvertToArea(
                   NewLine(NewPointLatLon(-76.5,42.45),
                   NewPointLatLon(-76.39,42.2),
                   NewPointLatLon(-76.2,42.9)
                           )
                   )
                 , NewPointLatLon(
                      [empty_dem].[Longitude (I)],
                      [empty_dem].[latitude (i)])
                 )
 )
SET [height (i)] = 1000
```

You can also update a raster by the values inside of an overlapping vector feature. In this example, we are updating the [Dem] with the [Area (I)] value contained in the [park] layer.:

```
OPTIONS CoordSys("dem" AS Component);
UPDATE
(
  SELECT empty_dem.*, park.[Area (i)] AS pArea
  FROM empty_dem, parks
  WHERE touches(parks.[geom (i)],
                NewPointLatLon([empty_dem].[Longitude (I)],
                               [empty_dem].[latitude (i)])
           )
  AND park_no = 1
 )
SET [Height (i)] = pArea
```

Generate a TIN from point data

Using a set of points, we can generate Triangulated Irregular Networks (TIN) as:

```
SELECT * FROM
 (
  SELECT Triangulation(AllCoords(elevpts.[geom (i)]))
  FROM elevpts
 )
```

The above example returned polygons. We can also return lines as well:

```
SELECT * FROM
 (
  SELECT TriangulationLines(AllCoords(elevpts.[geom (i)]))
  FROM elevpts
 )
```

Kriging from point data

The Kriging function in SQL determines the interpolated value for a location (i.e. a geometry or X,Y pair). However, raster data may be updated by replacing the [Height (I)] value with an interpolated value obtained by kriging for each location in a raster surface. In this case, we need an empty raster surface (or any raster surface if you are willing to overwrite the existing values) that we wish to repopulate all the pixel values with using a kriged interpolated value.

The syntax for using the kriging function is:

```
Kriging([drawing with the elevations], [elevation column],
[neighbors to include], [geometry to update])
```

And the query to update a new surface is:

```
OPTIONS CoordSys("empty_dem" AS Component);
UPDATE [empty_dem]
Set [Height (I)] = kriging(elevpts,[Height],-1,
NewPointLatLon([longitude (i)],[latitude (i)]))
```

Other adaptations of the kriging function allow users more flexibility such as the ability to enter the variogram model type (spherical or gaussian). In this case, the syntax is:

```
Kriging([drawing with the elevations], [elevation column],
[neighbors to include], [variogram model], [whether or not to
include voronoi neighbors], [geometry to update])
```

And the query to update the new surface with the spherical variogram model is:

```
OPTIONS CoordSys("empty_dem" AS Component);
UPDATE [empty_dem]
Set [Height (I)] = kriging(elevpts,[Height],-1,"spherical",true,
NewPointLatLon([longitude (i)],[latitude (i)]))
```

Generate contour data from points

Contour line generation in Manifold GIS is a function of the GUI, and therefore not built into the SQL engine.

Generate contour data from raster

Contour line generation in Manifold GIS is a function of the GUI, and therefore not built into the SQL engine.

Data Transformation - mathematical transformation of raster data

Mathematical transformation of raster data in Manifold GIS is a function of the GUI, and therefore not built into the SQL engine.

Projection definition and coordinate transformation

Geometries may be projected and defined on-the-fly with SQL in Manifold GIS. The following query projects the geometry from the parcels layer stored as *State Plane, NY Central* to the *UTM Zone 18* coordinate system obtained from the [dem] layer.

Change Projection

```
SELECT Project(Geom([geom (i)]), CoordSys("dem" AS COMPONENT))
FROM [parcels];
```

If you don't have another layer with a coordinate system built in, you can explicitly tell the SQL engine what coordinate system you want it in, like Latitude and Longitude:

```
SELECT Project(Geom([geom (i)]),CoordSys("Latitude /
Longitude")) AS g
FROM [parcels];
```

or, for UTM Zone 18N:

```
SELECT Project(Geom([geom (i)]),CoordSys("Universal Transverse
Mercator - Zone 18 (N)"))  AS g
FROM [parcels];
```

or for State Plane New York Central (1983 feet)[6]:

```
SELECT Project(Geom([geom (i)]),CoordSys("State Plane - New York
Central*(NAD 83, feet)")) AS g
FROM [parcels];
```

Define Projection

Finally, if the geometry is in the correct numerical format, but does not have a coordinate system assigned, you can assign a coordinate system. For example, assume that the parcel layer does not have a coordinate system defined, but should actually be UTM 18N. The query would be:

```
SELECT AssignCoordSys(Geom([geom (i)]),CoordSys("Universal
Transverse Mercator - Zone 18 (N)")) AS g
FROM [parcels];
```

[6] You must correctly spell out the coordinate system. In the State Plane example, notice the asterisk (*) - you must include this. If you include a space instead, the projection will fail.

Vector overlay - polygon in polygon overlay; point in polygon overlay; line in polygon overlay

Overlaying and finding geometry features contained within polygons are the same whether using points, lines, or polygons, and would take the form of:

```
SELECT parcels.*
FROM firm, parcels
WHERE contains(firm.[Geom (i)],parcels.[Geom (i)])
AND firm.zone = "AE"
```

in this case, it does not matter whether the parcels are points, lines, or polygons. In addition to contains, other operations may be used and include `adjacent`, `touches`, or `intersects`.

Raster processing - mathematical operations on one raster (sine, cosine, exponent)

Raster functionality in SQL supports virtually any mathematical function you would find in a spreadsheet application. Therefore, individual pixel values can be updated with the mathematical function as:

Sine

```
UPDATE [dem]
SET [Height (I)] = sin([Height (I)])
```

Cosine

```
UPDATE [dem]
SET [Height (I)] = cos([Height (I)])
```

Convert Degrees to Radians

```
UPDATE [dem]
SET [Height (I)] = deg2Rad([Height (I)])
```

Exponent

```
UPDATE [dem]
SET [Height (I)] = [Height (I)]^2
```

Raster processing - mathematical operations on two rasters - adding, subtracting, minimum, maximum

Manifold's raster processing is typically reserved for the GUI. However, there are some ways to force it to perform raster processing for combining multiple raster layers. To do this, the centerpoint of each pixel in one raster layer is treated as a vector object and turned into an individual area feature. The centerpoint of the second raster layer is left as a point, and a SELECT touches query is issued[7].

[7] Because you cannot guarantee that two raster layers will line up with one another, the pixel objects must be turned into geometries to perform a spatial comparison.

Neighborhood functions - average, minimum, maximum, most frequent

Neighborhood functions on raster surfaces are computed using the UPDATE command within an SQL query.

Average

```
UPDATE [dem]
SET [Height (I)] = HeightAvg([dem],NewPointLatLon
                                ([Longitude (I)],[Latitude (I)]))
```

Minimum

```
UPDATE [dem]
SET [Height (I)] = HeightMin([dem],NewPointLatLon
                                 ([Longitude (I)],[Latitude(I)]))
```

Maximum

```
UPDATE [dem]
SET [Height (I)] = HeightMax([dem],NewPointLatLon
                               ([Longitude (I)],[Latitude (I)]))
```

Most Frequent

```
UPDATE [dem]
SET [Height (I)] = HeightMaj([dem],NewPointLatLon
                               ([Longitude (I)],[Latitude (I)]))
```

Statistical functions - calculating areas, perimeters and lengths

Descriptive statistics on geometries can be calculated to include area, length, or perimeter. If the data is in a particular coordinate system, you can specify the units.

```
SELECT area([geom (i)],"ft") AS ParcelArea_SqFt,
        area([geom (i)],"m") AS ParcelArea_SqMeters,
        Length([geom (i)],"ft") AS ParcelLength
FROM parcels
```

Cross tabulation of two data categories

Cross tabulation matrices are created by using the TRANSFORM function in conjunction with a PIVOT table. In this example, we are summing the area of each park ([parks]) that intersects a flood polygon ([firm]), and cross tabulating it with the actual flood zone. Also, the small calculation in the first line returns the area in square feet, and then divides that by 43,560 to convert the area to acres:

```
TRANSFORM Round(SUM(Area(g,"ft")/43560),2)
SELECT name
FROM
     (SELECT ClipIntersect([parks].[Geom (I)],
      [firm].[Geom (I)]) AS g, [name],[zone]
      FROM [parks],[firm]
      WHERE name <> ""
      AND touches([parks].[Geom (I)],[firm].[Geom (I)])
     )
GROUP BY name
PIVOT zone
```

General - specify distance buffers; polygons within a distance of selected buffers; find nearest features

The ability to specify distance buffers was already addressed in a previous section. However, finding polygons within a specified distance and finding nearest features are calculated as:

Polygons within a distance

```
SELECT parcels.*
FROM  parcels, parks
WHERE distance(parcels.[geom (i)], parks.[geom (i)],"ft") < 800
```

Nearest Features

This query finds the closest road from each tree - however, to speed the query up, a distance threshold of 200 ft. is used as a cutoff value.

```
 SELECT min(dist) AS minium_distance,
        first(roadid) as roadid, site_id
FROM
    (SELECT roadid, site_id,
     distance(trees.[Geom (I)],roads.[Geom (I)]) AS dist
     FROM trees, roads
     WHERE
      distance(trees.[Geom (I)],roads.[Geom (I)],"ft") < 200
     ORDER BY dist
  )
 GROUP BY site_id
```

3D analysis - generating slope and aspect

Slope and aspect require a raster layer and a location for where the user wants the slope and height to calculated. The location is returned as a point geometry.

Slope

```
OPTIONS CoordSys("dem" AS component);
UPDATE [empty_dem]
SET [Height (I)] = SlopeHeight([dem],
                   NewPointLatLon([Longitude (I)],[Latitude
(I)]))
```

Aspect

```
OPTIONS CoordSys("dem" AS component);
UPDATE [empty_dem]
SET [Height (I)] = AspectHeight([dem],
                   NewPointLatLon([Longitude (I)],[Latitude
(I)]))
```

Identifying watersheds

Watersheds are delineated in the GUI, and not supported in Manifold 8 SQL.

Network functions - choosing the optimal path through a network

Network functions are computed in the GUI, and not supported in Manifold 8 SQL.

Defining a drive-time zone

Drive-time functions are computed in the GUI, and not supported in Manifold 8 SQL.

Geocoding addresses

You can create geometries from addresses in SQL using a hand-typed address, passing a field or even a concatenation of multiple fields:

```
VALUES (LocateAddress("16 Bush Lane, Ithaca, NY"))

SELECT LocateAddress([addrno] & [addrstreet] & " Ithaca, NY")
FROM parcels
```

In addition, you can return the X,Y values of a geocoded address:

```
SELECT
  LocateAddressLon([addrno] & [addrstreet] & " Ithaca, NY"),
  LocateAddressLat([addrno] & [addrstreet] & " Ithaca, NY")
FROM parcels
```

Topological intersection[8]

SQL in Manifold supports all of the classical topological intersection methods (intersect, union, identity) using variations of the clipintersect and clipsubtract methods. These capabilities are more sophisticated than most SQL presented in this book. Nonetheless, the command builds upon the more foundational SQL that you have already seen. The parcel and parks layers are used as illustrative examples:

<u>Intersect</u>

The basic principle for intersection is to clip two area layers using the ClipIntersect function, and then joining the subsequent data tables to the clipped features.

```
SELECT * FROM
  (
   SELECT ClipIntersect(parcels.id,parks.id) AS g,
          int(parcels.id) AS parcelid, int(parks.id) AS parkid
   FROM parcels, parks
   WHERE touches(parcels.id,parks.id)
  )
RIGHT JOIN [parcels] ON parcels.id = parcelid
RIGHT JOIN [parks] ON parks.id = parkid
```

[8] for these examples, we are using the id field to obtain the geometry rather than the [geom (i)] field so as to provide less clutter on the page.

Union

Union is slightly more complicated and requires the joining of three separate queries: clipping the two layers together with ClipIntersect, obtaining the parts of layer 1 that are not in layer 2 using ClipSubtract, and obtaining the parts of layer 2 that are not in layer 1. After the geometries are assembled, the attribute tables are joined in.

```
SELECT * FROM
(
  SELECT ClipIntersect(parcels.id,parks.id) AS g,
         parcels.id AS cid, parks.id AS rid
  FROM   parcels, parks
  WHERE  touches(parcels.id,parks.id)

UNION ALL
 SELECT Clipsubtract(parcels.[Geom (I)],
        (SELECT UnionAll(id) FROM parks)) AS g,
         parcels.id AS cid, 0 AS rid
         FROM parcels
         WHERE touches(parcels.id,parks.id)
UNION ALL
 SELECT ClipSubtract(parks.[Geom (I)],
        (SELECT unionall(id) FROM parcels)) AS g,
         0 AS cid, parks.id AS rid
         FROM parks
         WHERE touches(parcels.id,parks.id)
)
LEFT JOIN [parcels] ON parcels.id = cid
LEFT JOIN [parks] ON parks.id = rid
WHERE IsArea(g)
```

Identity

Identity requires two separate queries: clipping the two layers together, and then finding those geometries in layer 2 that are not layer 1. After the geometries are reassembled, the attribute tables are joined in.

```
SELECT * FROM
(
 SELECT
  ClipIntersect(parcels.id,parks.id) g, int(parcels.id) AS cid,
  int(parks.id) AS rid
 FROM parcels, parks
 WHERE touches(parcels.id,parks.id)

UNION ALL

SELECT
 ClipSubtract(parks.id,parcels.id) AS g, int(parcels.id) AS cid,
 int(parks.id) AS rid
FROM parcels, parks
WHERE touches(parcels.id,parks.id)
)
RIGHT JOIN [parcels] ON parcels.id = cid
RIGHT JOIN [parks] ON parks.id = rid
WHERE IsArea(g)
```